03736216 L2 (2)

FORESTRY COMMISS

CW01651165

Roe Deer Biology and Management

P. R. Ratcliffe and
B. A. Mayle
Wildlife & Conservation Research Branch,
Forestry Authority

LONDON: HMSO

First published 1992

ISBN 0 11 710310 1
FDC 149.6 Capreolus capreolus:15:451.4:(410)

KEYWORDS: Forestry,
Protection, Wildlife management

Enquiries relating to this publication
should be addressed to:
The Technical Publications Officer,
The Forestry Authority, Forest Research Station,
Alice Holt Lodge, Wrecclesham,
Farnham, Surrey, GU10 4LH.

Front cover: Roe deer in winter.
(© *Hellio and Van Ingen/NHPA*)

Contents

Roe Deer Biology and Management

Summary

Roe deer are distributed widely in Great Britain and are managed for a number of reasons including the reduction of impact on trees and vegetation and their exploitation as a game species. Population data, especially on survivorship, on which to base management plans, are difficult to obtain and have previously been unavailable.

This Bulletin gives a brief account of roe deer biology, insofar as this affects their management. It proposes a management strategy and methodology, based on deer numbers, population dynamics and habitat changes, and measurement of their impact on trees and other vegetation. Examples of population models are presented to illustrate the considerable regulatory effect of juvenile mortality on roe deer populations. The high levels of juvenile mortality appear to have been under-estimated previously, and consequently culling levels aimed at preventing population increases have been over-estimated. It appears that culls of the order of 15–25% will prevent many British roe deer populations increasing in size.

Biologie et Contrôle des Chevreuils

Résumé

Les chevreuils vivent partout en Grande-Bretagne et il est nécessaire de contrôler leurs effectifs pour plusieurs raisons: d'abord pour réduire leur impact sur les arbres et la végétation en ensuite pour permettre leur exploitation en tant que gibier. Des données spécifiques concernant leurs effectifs et particulièrement leur capacité de survie – dans le but d'établir des programmes de contrôle – sont difficiles à obtenir; par le passé, il a même été impossible de se procurer ce type de renseignements.

Ce Bulletin donne un bref aperçu des caractéristiques biologiques des chevreuils, dans la mesure où cela affecte leur contrôle. Il propose une stratégie et une méthodologie de contrôle basées sur leurs effectifs, leur dynamique de groupe et leurs changements d'habitat, et la mesure de leur impact sur les arbres et la végétation en général. Des modèles d'évolution de leurs effectifs sont présentés de manière à illustrer l'effet régulateur exercé par la mortalité des jeunes sur l'effectif global des chevreuils. Il semble que les taux élevés de mortalité des jeunes ont été jusqu'à présent sous-estimés; par conséquent, le nombre d'animaux éliminés afin d'empêcher une trop forte croissance des effectifs a été surestimé. A l'avenir, un pourcentage d'élimination de l'ordre de 15 à 25% empêchera la multiplication du nombre des chevreuils en Grande-Bretagne.

Rehwildbiologie und Kontrolle

Zusammenfassung

Rehwild ist in Grossbritannien weit verbreitet und wird aus mehreren Gründen kontrolliert, darunter u.a. zur Beschränkung seiner Auswirkung auf Bäume und andere Vegetation und zur Nutzung als Jagdwild. Bestandsdaten, insbesondere Überlebensraten, auf die sich Kontrollpläne basieren, sind schwer zu erlangen und waren bisher unerhältlich.

Dieses Bulletin gibt einen kurzen Umriss der Rehwildbiologie insofern sie seine Kontrolle beeinflusst. Es schlägt Kontrollstrategie und -methoden vor die sich auf Rehbestand, Populationsdynamik, Standortveränderungen und Messung der Auswirkungen auf Bäume und andere Vegetation gründen. Beispiele von Populationsmodellen werden gegeben um den beachtlichen Effekt der Jungtiersterblichkeit zu illustrieren. Die hohen Sterberaten von Jungtieren sind anscheinlich bisher unterschätzt worden, und folglich wurden Abschussraten zur Verhütung einer Zunahme des Bestandes zu hoch angelegt. Es scheint, daß Abschussraten von ungefähr 15–25% das Vergrössern vieler britischer Rehwildbestände verhindern werden.

Roe Deer Biology and Management

Introduction

Roe deer (*Capreolus capreolus*) is our most widely distributed deer species, occurring throughout Great Britain except for most of Wales and the English Midlands (Figure 1). They are native to Great Britain, but not Ireland, and had been virtually restricted to a few parts of Scotland by the 18th century. Since then a natural spread related to the expansion of forestry and a number of introductions, particularly in southern England, have led to a large increase in their range. Roe deer are sympatric (occupying the same area) with red (*Cervus elaphus*), sika (*Cervus nippon*), fallow (*Dama dama*) and muntjac (*Muntiacus reevesi*), in different parts of their range in Great Britain.

This Bulletin presents a scientific basis for managing deer populations. It involves the collection of information on population dynamics (the basis for determining the rate of increase or decrease of the population) and the estimation of density to calculate population size. Culls are based upon these data and the results monitored by retrospective checks on population size using cohort analysis.

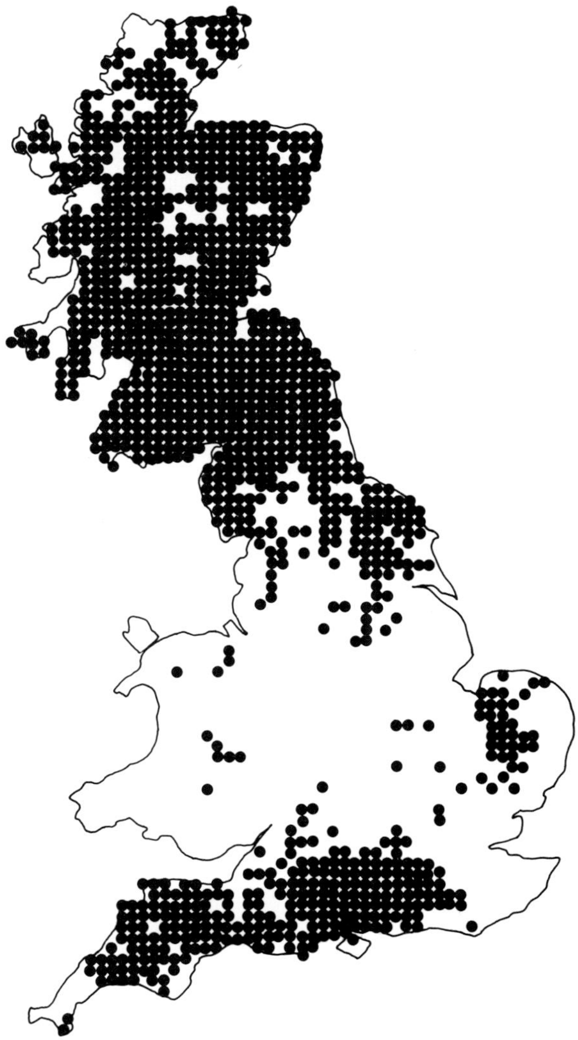

Figure 1. Distribution of roe deer in Great Britain.

Why manage deer?

Due to a lack of deer predators in Great Britain most unmanaged deer populations will increase in numbers until food shortage or social behaviour triggers regulatory mechanisms causing a balance between mortality and emigration, and natality and immigration. Unfortunately this balance is not reached before heavy impact on the habitat and a decline in condition of the deer have occurred. A few deer populations are declining, mainly due to a combination of habitat destruction and over-exploitation.

Deer and their habitats may be managed for at least six reasons.

1. To protect timber and agricultural crops from damage.
2. To conserve plant assemblages and their dependent animals.
3. To exploit the deer for trophy hunting and venison production.
4. To enhance the aesthetic appeal of the environment and to provide recreational opportunities for people.
5. To conserve deer populations, as a species in their own right.
6. To regulate population size and prevent starvation.

The need to manage deer to prevent starvation depends upon the ethical principle that, in the absence of predators, man is responsible for regulating deer numbers. Such treatment of deer is, however, inconsistent with that of most other species of British mammals and birds.

Whether the presence of animals in an area manifests itself as a pest problem or an asset, depends upon land use practice, as well as population size and the suitability of available habitats. The presence of deer must be recognised as an aesthetic benefit (their very presence giving pleasure to people), an exploitable resource (recreational hunting and venison), and as a problem (modification of habitats and damage to trees and other crops). These factors are not mutually exclusive. How deer are perceived will depend on land use objectives and recreational demands on the countryside. Management must consider all of these possible objectives.

Many deer populations are increasing in numbers and control is not being achieved. However, some live in fragile and fragmented habitats and over-exploitation could cause serious declines, e.g. red deer on Exmoor. The particular situation must be carefully appraised locally before constructing a deer management plan. In British forests there is seldom justification for not managing deer populations in some way.

Deer management strategy

Deer do not adhere to geographical units of particular land ownership. Therefore, landowners and managers occupying continuous areas of deer range must co-operate. If the objectives of adjacent land managers vary, then there is likely to be a greater conflict of interests than, for example, if all of the land is being managed for one purpose. Whatever the land use objectives, a more workable deer management strategy is most likely to emanate from a forum of open discussions, and attempts to understand a neighbour's point of view. The organisation of a deer management group (DMG), as applies to red deer populations throughout most of their range in Scotland, is a valuable means to this end.

Deer management requires the application of art and science. The art involves the use of field skills and the development of local knowledge; the fundamentals of this can be taught by skilled practitioners or learned from books, but there is no substitute for experience. The science is more easily learned, but requires a firm commitment and an objective approach. The management of deer requires an assessment of:

1. Impact (p.4)
2. Density (p.5)
3. Predicted changes in habitat structure (p.10)
4. Population dynamics (p.12)
5. Previous culls (p.22)

The management strategy should be decided upon using the flow-chart (Figure 2). Note the clear distinction between controlling numbers by overall population reduction and controlling impact by localised shooting in and around vulnerable areas.

Management must be predictive. Reacting to deer damage by increased shooting is not an ideal way to manage a deer population. The essence of predictive wildlife management is in the ability to forecast changes in habitats. Deer use different components of the habitat differentially, spending more time in some types of forest than others. For example, the pre-thicket and early thicket stages generally offer an optimum mixture of food and cover and therefore support higher numbers than other structural types. The prediction of changes in forest structure enables us to anticipate changes in carrying capacity, and modify management (culling

Determine: Impact on habitat/crop
Predicted changes in forest structure
Population features: Population dynamics
Density
Previous culls

Evaluate:

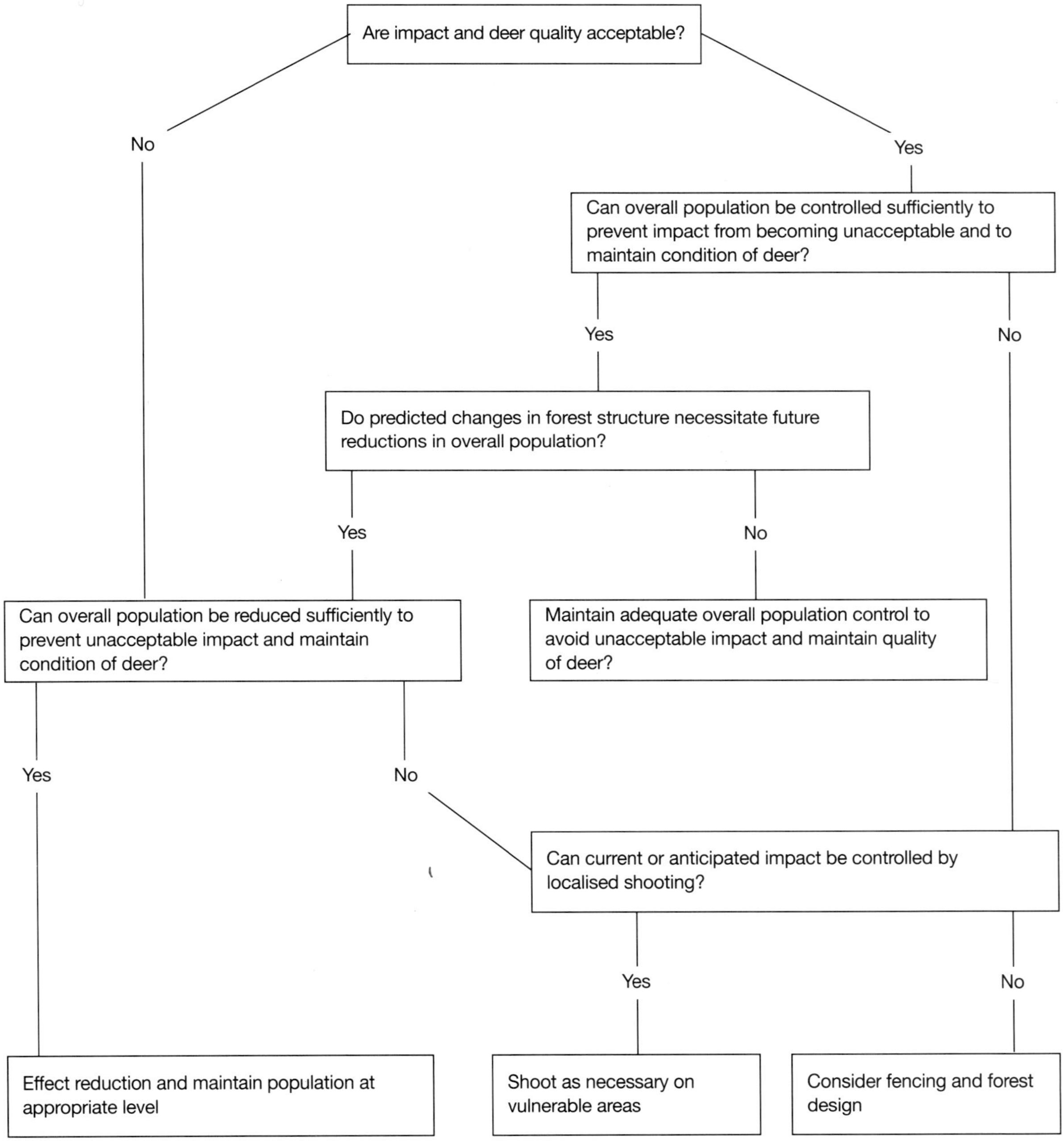

Figure 2. Roe deer management strategy.

requirements and habitat modifications) accordingly.

Tree damage and vegetation impact

Roe deer browsing on commercially important tree crops can reduce early height increment, thereby increasing establishment costs by increasing the period for which weeding is necessary. The economic significance of current browsing damage over the ensuing crop rotation is difficult to predict and is probably dependent upon the capability of the tree to recover, which is in turn dependent on site quality. However, heavy browsing of vulnerable species (e.g. Norway spruce, *Picea abies*; Douglas fir, *Pseudotsuga menziesii*; silver firs, *Abies* spp.; and most broadleaves) certainly causes a delay in trees reaching maturity and is clearly important. Browsing may also result in trees producing multiple leading shoots, which can cause economic loss by reducing timber quality. The extent and significance of early browsing damage can be assessed by placing individual tree guards around a sample of newly planted trees. Subsequent comparisons of the height of protected and unprotected trees will indicate the ability of trees to recover from browsing, and any delay in reaching the upper height which is normally browsed by roe deer (approx. 1 m) will give some impression of the ability of damaged trees to effect compensatory growth, though this may continue to occur throughout the rotation.

Fraying and thrashing with antlers is usually localised, but can cause significant damage on valuable specimen trees and small areas of broadleaves. Roe deer seldom strip the bark from trees with their teeth.

The selective browsing of tree seedlings and herbaceous plants can result in considerable modification to plant assemblages at relatively low densities of deer (<5 km^{-2}). This may be of importance in the management of vegetation, and its dependent animals, for conservation objectives.

Objective assessments of the amount of deer damage occurring on trees can be made following the procedures provided in Forestry Commission Leaflet 82 *Assessment of wildlife damage in forests* (Melville, Tee and Reynolds, 1983).

Protection by physical barriers

It may be necessary to use tree guards (Forestry Commission Arboricultural Leaflet 10 *Individual tree protection* (Pepper, Rowe and Tee, 1985)) or fencing (Forestry Commission Bulletin 102 *Forest fencing* (Pepper, 1992)) to protect trees or vegetation from damage. Treeshelters (Forestry Commission Handbook 7 *Treeshelters* (Potter, 1991)) may also offer some protection along with the added benefits of enhanced growth. However, due to the variability in degradation of treeshelters, trees are continually becoming exposed to deer as fraying stocks. This may necessitate continuing protection with tree guards and confirms the undesirability of premature removal of treeshelters in some areas.

Roe deer biology and social structure

Roe deer are largely solitary or occur in small family groups. Adult does, once established, will stay on their own home range for several years. Young females form their own home ranges in their second or third year, often overlapping with that of their mother, or they may emigrate, usually when one year old. The trigger which causes dispersal and regulation of females is unknown. Female home ranges are not defended and overlap with other females and males.

Adult bucks may obtain territories, usually by successfully competing with existing territory holders, or may fail to achieve this, remaining as non-territorial bucks. The territory is usually larger than that of a doe's home range, and frequently overlaps the ranges of several does. Bucks are territorial from April to August and become more tolerant of one another soon after the rut. During severe weather, territoriality may break down and all deer may share communal feeding grounds, but these are usually outside territorial boundaries. Territory and home range size vary according to habitat but defended territories seldom exceed 60 ha.

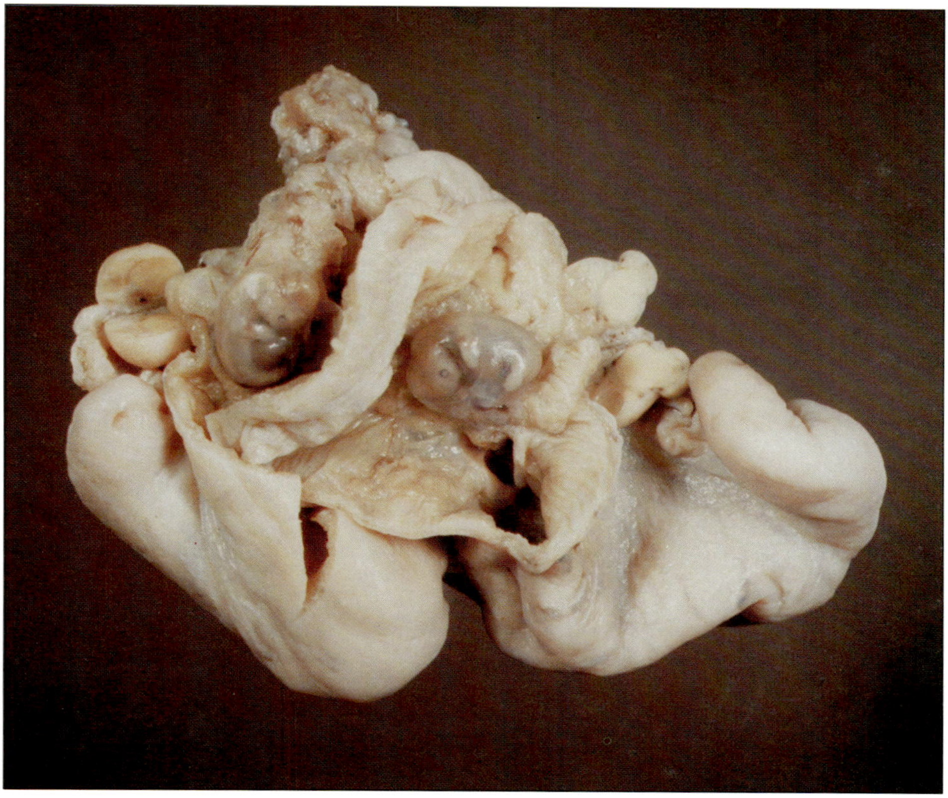

Plate 1. *Uterus and ovaries of roe doe cut open to show developing embryos and* corpus luteum *in the ovary.*

Plate 2. *Ovaries of roe doe sectioned to show developing follicles (A) and yellow* corpora lutea *(B). (40065)*

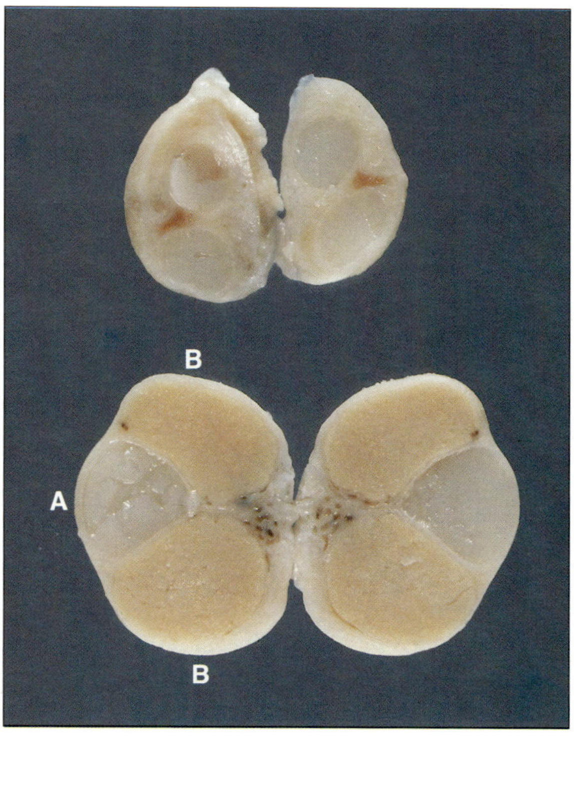

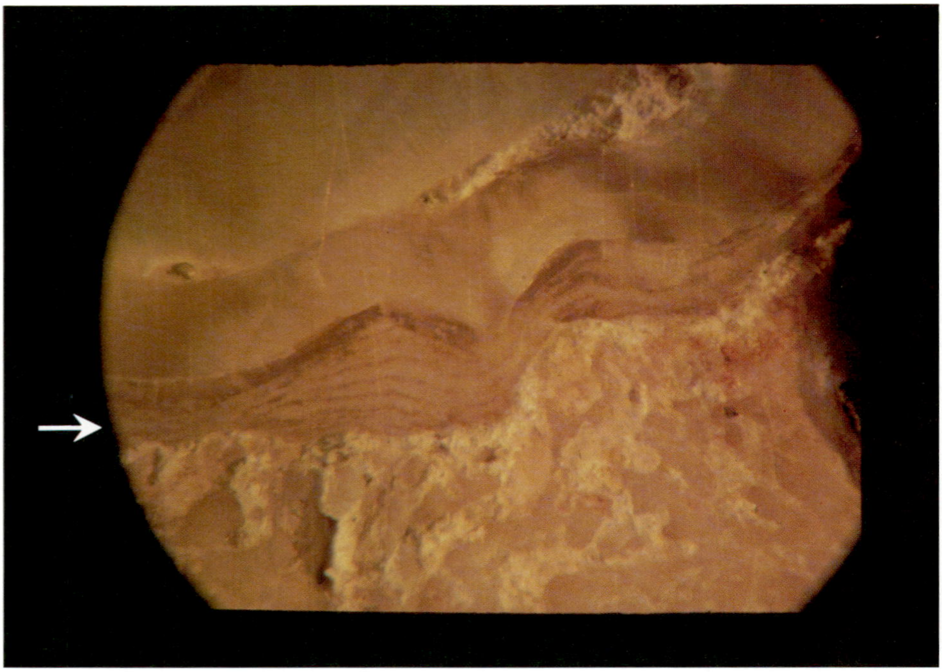

Plate 3. *Sectioned tooth showing cementum layers.*

Plate 4. *Infant kid exposed to heavy rain.*

The ranges of non-territorial males can be twice the size of territories, and may overlap with those of several bucks. Territories are stable from year to year being continuously occupied by the same or by replacement bucks, while the ranges of non-territorial males change with animals emigrating in the spring of the second or third year if they are unable to acquire a vacant area locally. Territorial bucks seldom hold territories for more than 3 years. Territorial boundaries frequently coincide with natural features such as streams, woodland rides or roads.

Roe deer rut in late July or early August but kids are not born until the following May or June. This apparently long gestation period is due to embryonic diapause (delayed implantation) of the fertilised egg which implants into the uterine wall in early January. Gestation then proceeds normally. At about two-thirds of the way through pregnancy, around the end of March or early April, the foetus has almost achieved its final body length and has a fully developed coat of hair with spots. This sometimes leads to the erroneous interpretation that the foetus is almost at full term. However, the internal organs are not fully developed at this stage and the final birth weight of 1–2 kg will only be achieved much later, around mid May to June.

The 'false rut' is an apparent resurgence of sexual activity by some deer in and around October, but its significance is not fully understood. It may simply explain late rutting caused by females returning to oestrus (heat) following earlier missed or unsuccessful matings. Does often return to the same place to give birth in consecutive years. Females normally breed first at 14 months of age producing their first kids around their second birthday but, in some populations, some kids ovulate and become pregnant at 3-4 months of age.

Assessing the number of deer

Roe deer populations inevitably occupy a range which includes a rather large proportion of dense cover in which deer cannot be seen. The estimation of numbers in these habitats must either rely on indirect methods (i.e. dung counts) or on direct methods conducted over long time periods which allow a high probability that deer will move and be seen during the count periods (i.e. vantage point counts).

Dung counts

Dung (or pellet-group) counting is the most useful indirect method. A major advantage is that dung counting provides an estimate of the average deer density over a period of several months, which is more useful to the deer manager than an estimate, or indeed a census, representing a single instant in time.

The principle of the dung count method is simple. It is based on the assumption that the more deer in an area the higher the density of dung to be found on the ground, and that this relationship will follow a linear form. The main factors, other than the number of deer, influencing the number of dung groups to be found, are the defecation rate of the deer (number of defecations each day), and the decay rate (time taken for dung groups to decompose or disappear). Clearing permanent plots and counting dung groups after a period will eliminate the need to assess decay rates, but in practice this method only works well at exceptionally high densities of deer and it is much more labour intensive than the 'single-visit' method described here. The defecation rate for roe deer is about 17–23 groups per day and this varies little between different habitats or geographical areas (Mitchell *et al.*, 1985). In the absence of local information a value of 20 groups per day can be assumed as an average.

Decay rates will vary between locations and these must be determined locally in a variety of habitats. To determine decay rates fresh dung must be observed over a period. Fresh dung groups can be located after watching animals defecating or they can be obtained directly from the rectum of culled deer. If deer have been observed defecating, dung groups can be left *in situ* and simply marked with a small stake or cane. Alternatively they can be collected and placed in a particular habitat type and again labelled or marked with a stake. Dung counts are most effective in the spring and therefore dung decay should be determined over winter, with decay plots being set up in the autumn. The date on which the dung groups were observed to be fresh should be recorded and observations on a monthly basis should be made to determine when the group has decayed completely (a dung group should initially include more than 40 pellets with decay being considered complete when fewer than six pellets remain visible). About six dung groups should be observed in each habitat available. Decay may only take a few months in rich broadleaved habitats and up to 15 months or more in coniferous acidic ones.

Dung counts must be related to distinct habitats used differentially by the deer. The proportion of early successional stages present in the forest will influence the availability of food and shelter for roe deer. The particular habitat types selected will depend upon local circumstances, and should reflect local knowledge of deer activity and usage. For example in Alice Holt Forest, Hampshire, a very diverse forest, 12 habitat types were recognised on the basis of age class of trees and vegetation types, which would probably be differentially used by the deer. These were as follows:

Conifer establishment
: Conifers 0–5 years old. Mainly Norway spruce (*Picea abies*) and Corsican pine (*Pinus nigra*) less than 1 m high. Ground flora of hazel (*Corylus avellana*), bracken (*Pteridium aquilinum*), bramble (*Rubus fruticosus*) and wavy hair grass (*Deschampsia flexuosa*).

Conifer pre-thicket
: Conifers, mainly Corsican pine 5–10 years old. Ground flora of bramble, hazel and dense bracken.

Conifer thicket
: Conifers, mainly Corsican pine, Norway spruce, hybrid larch (*Larix eurolepis*) and western hemlock (*Tsuga heterophylla*), aged 10–20 years old. Ground flora of bracken, wavy hair grass and raspberry (*Rubus idaeus*).

Conifer pole-stage
: Conifers aged 20–30 years old, mainly Norway spruce, western hemlock, Corsican pine, hybrid larch and Douglas fir (*Pseudotsuga menziesii*). Ground floor of bracken, bramble, wavy hair grass and dead needles.

Conifer pre-felling
: Conifers more than 30 years old. Corsican pine, western hemlock, Norway spruce, Japanese larch (*Larix kaempferi*). Canopy closed in most places with a reduced ground flora. Scattered patches of short bracken and wavy hair grass.

Broadleaved establishment
: Broadleaved trees aged 0–3 years. Mainly oak about 1 m high. Ground flora of bramble, hazel, bracken, holly (*Ilex aquifolium*) and Yorkshire fog (*Holcus lanatus*).

Broadleaved pre-thicket
: Broadleaved trees of 4–15 years. About 2.5 m high. Mainly oak interspersed with regenerating hazel. Ground flora of Yorkshire fog, bramble, bilberry (*Vaccinium myrtillus*), willowherb (*Chamaenerion angustifolium*), dog's mercury (*Mercurialis perennis*) and bracken.

Broadleaved thicket
: Broadleaved trees of 15–25 years. A mixture of oak (*Quercus* spp.) with some Norway spruce and hazel. Profuse shrub vegetation with dog's mercury, bramble, wild strawberry (*Fragaria vesca*), ground ivy (*Glechoma hederacea*) and willowherb.

Broadleaved pole-stage
: Broadleaved trees of 25–40 years. A mixture of oak, alder (*Alnus glutinosa*) and beech (*Fagus sylvatica*). Canopy and understorey species of holly, hawthorn (*Crataegus* spp.), yew (*Taxus baccata*) and hazel. Ground flora of bramble, dog's mercury and creeping soft grass (*Holcus mollis*).

Broadleaved pre-felling
: Broadleaved trees older than 40 years. A mixture of oak, ash (*Fraxinus excelsior*), sweet chestnut (*Castanea sativa*) and western red cedar (*Thuja plicata*). Canopy and understorey species of hawthorn, yew and hazel. Ground flora of bramble, honeysuckle (*Lonicera periclymenum*), stinging nettle (*Urtica dioica*), wavy hair grass and bracken.

Recently felled broadleaved woodland
: Originally oak woodland with hawthorn, hazel and field maple (*Acer campestre*). Newly growing coppiced stumps and ground flora including bramble, raspberry and bracken.

Pasture

A semi-permanent grassland used for hay and grazing. Ground vegetation of dock (*Rumex* spp.), creeping thistle (*Cirsium arvense*), meadow buttercup (*Ranunculus acris*), yarrow (*Achillea millefolium*) and grasses (Gramineae).

It is quite possible that the stratification of habitats does not relate exactly to the way deer are actually using an area. This will become obvious after the dung count has been completed and it may be possible to combine two or more habitat types if there is little difference in the deer usage of them. Alternatively, a very wide variation in dung counts between plots, within a habitat, may indicate that further stratification is required. For example, high dung counts close to edges or open patches may indicate that more precision would be achieved by stratifying the edge or open space as a separate habitat. Within each habitat or structural type the procedure is as follows.

Habitats are outlined on a stock map, and an easily locatable landmark close to the perimeter, such as a road end or stream intersection, is selected as a starting point. A straight line drawn from the starting point, through the selected area, serves as a ground survey line (Figure 3). This should avoid running parallel to edges, roads, rides, etc. The starting point is located on site and the map bearing is transferred to an orienteering compass. A pre-determined distance is walked depending upon the size of the habitat (50 or 100 m) along the compass bearing, and a 7 × 7 m plot is laid down with its axis along the survey line, using a tape measure for distances and the compass for judging right angles.

The compass bearing is followed for the same pre-determined distance and the procedure continued until eight plots have been assessed. (Using fewer than eight plots considerably reduces the confidence of the estimate, while using more than eight plots usually contributes little to increasing precision.)

Each plot is carefully and systematically

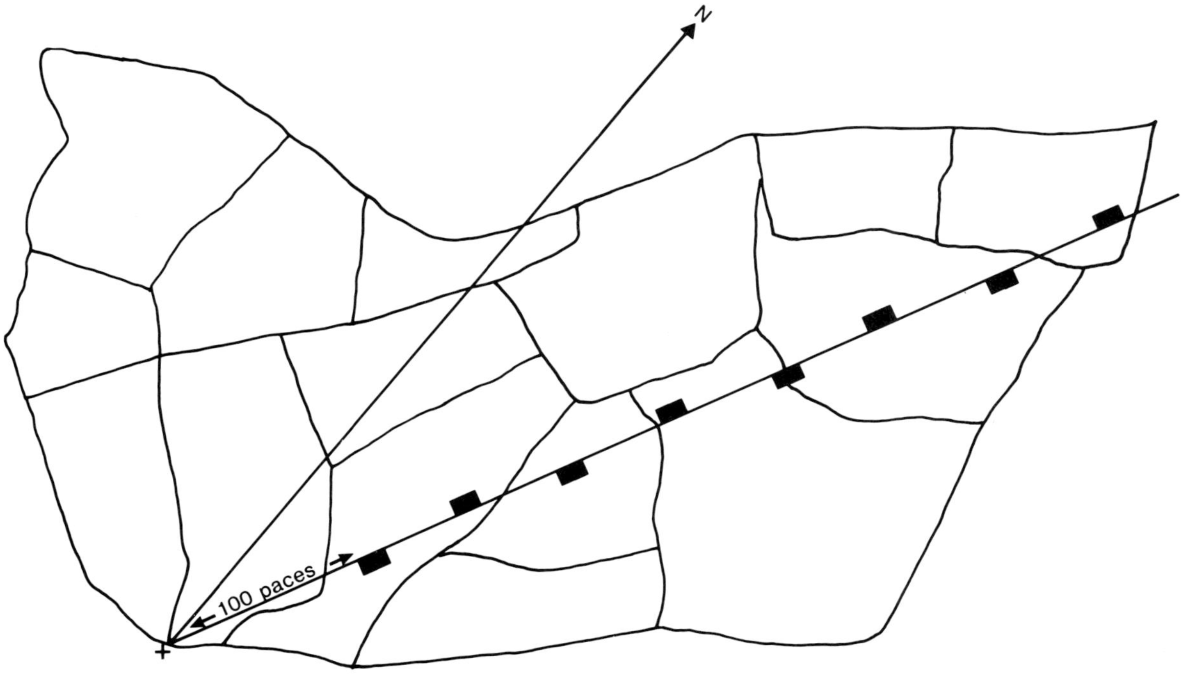

+ Starting point — Ground survey line

■ Plot ■——■ Distance between plots 50 paces

Figure 3. Survey and plot lay-out for dung assessment.

searched in bands of about 1 m wide for the presence of dung groups. It is important to search very carefully. It is almost certainly necessary to search on hands and knees in dense vegetation. The vegetation must be parted carefully and searching continued under tussocks. Dung may be hidden below recently fallen litter. Searching should penetrate the recently fallen and previous years litter but should *not* penetrate the litter/humus layer. A committed and methodical approach is essential. The number of dung groups is counted and recorded. A dung group is defined as a cluster of more than six pellets and usually dung groups are obvious large accumulations of individual pellets. Very low numbers of pellets can usually be associated with strings of pellets caused by the animal defecating while on the move or the pellets being scattered on hitting the ground. These are generally easily identified and assessed after careful searching. Occasionally, difficulties may be encountered in areas of high dung-group density when two or more groups have been deposited on the same spot. With experience it is always possible to separate these with confidence on the basis of individual pellet appearance, size and age. If dung groups are found on the edge of a plot they should be counted or rejected on the basis of how much lies inside or outside of the plot; only occasionally are dung groups found exactly on the edge, and in this case they should be alternately included and rejected.

Roe deer dung can be confused with that of other species of deer and with that of sheep or goats. It is important, therefore, to check standards periodically with other users of the technique and to check against pellets of known origin.

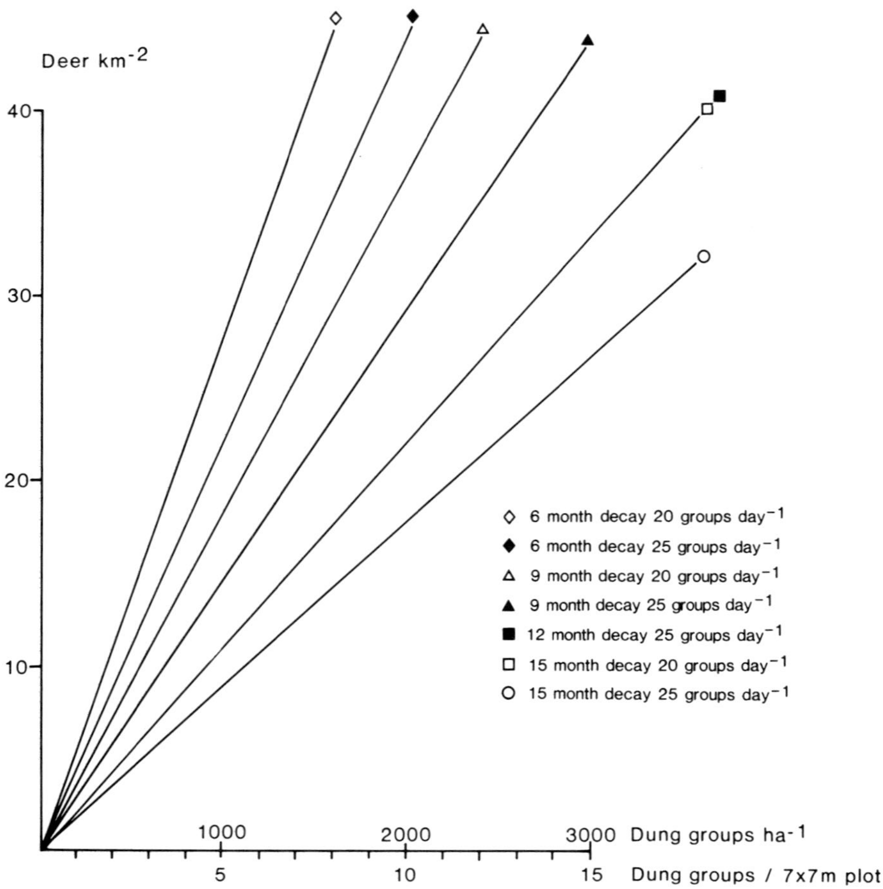

Figure 4. Relationships between dung and deer densities for a range of defecation and decay rates.

Dung count calculations

The results of a single dung count in a particular habitat might look like this:

Forest: KIELDER Location: COMPTS 401–414
Habitat type: Pre-thicket

Plot no.	1	2	3	4	5	6	7	8	Total	Mean
Dung gps.	4	1	4	5	3	7	4	3	31	3.875

The mean is calculated by dividing the total number, 31, by the number of plots, 8. This value of 3.875 can be rounded to the nearest whole number, 4.

The mean value of four dung groups can be simply applied to the graph (Figure 4) using the line appropriate to the particular conditions of defecation and decay rates to obtain deer density. For example, selecting the line for 9-month decay rate with a defecation rate of 20 groups day^{-1}, 4 dung groups gives a deer density of 15 deer km^{-2}.

However, there will of course be some error associated with this estimate and it is useful to have an understanding of the likely variation around the mean. This is usually expressed as confidence limits. This is most easily determined with the aid of a pocket calculator which can provide standard deviations. Additionally, 'Tables of *t*', will be required. Complete sets of these tables are available in most elementary statistics books, but the values most commonly needed for dung counts are provided in Table 1.

The procedure used to determine the confidence limits of the estimate is first to enter the individual counts of dung groups from each plot into the calculator, pressing the sum, or Σ, key after each value. For example,

4Σ, 1Σ, 4Σ, 5Σ, etc.

After all eight values have been input, depress the 'MEAN' key for the mean value of dung groups per plot, and the standard deviation key for the standard deviation. The standard deviation is a measure of the variation of individual counts (plots) around the mean. We must next convert this to standard error, a measure of the likely error associated with the mean of the whole population (of dung) which we are estimating. The standard error can be obtained by:

$$\frac{\text{Standard deviation}}{\sqrt{\text{Number of samples}}}$$

In the example, mean = 3.875; standard deviation = 1.61535; number of samples = 8. Therefore,

$$\frac{1.61535}{\sqrt{8}} = \frac{1.61535}{2.828} = 0.5712$$

We can now calculate the confidence limits of the estimate by multiplying the standard error by the appropriate value of *t* from Table 1. The probabilities of 80, 90 and 95% (Table 1) indicate that the estimate obtained will have this particular probability of being correct.

Table 1. Values of *t*.

	Probability (%)		
n -1	80	90	95
6	1.440	1.943	2.447
7	1.415	1.895	2.365
8	1.397	1.860	2.306
9	1.383	1.833	2.262
10	1.372	1.812	2.228
11	1.363	1.796	2.201
12	1.356	1.782	2.179
13	1.350	1.771	2.160
14	1.345	1.761	2.145
15	1.341	1.753	2.131

In our example we have a standard error of 0.5712, and the number of samples minus 1 (*n*-1 in Table 1) is 7. Thus:

$$0.5712 \times 1.415\ (80\%) = 0.81$$
$$0.5712 \times 1.895\ (90\%) = 1.08$$
$$0.5712 \times 2.365\ (95\%) = 1.35$$

These confidence limits indicate the probability of repeated estimates falling between these upper and lower values (i.e. 3.875 plus or minus the confidence limits). For example, we can be 80% certain of the real dung density falling between 3.07 and 4.69, 90% certain of it falling between 2.80 and 4.96 and 95% certain of it falling between 2.53 and 5.23.

Now these upper and lower values, as well as

the mean, can be applied to Figure 4 to give the range of deer densities, with 80, 90 and 95% confidence limits. It will be obvious now that the more widely the individual plot counts vary from one another, the greater the confidence limits become. It is also clear that as greater precision is demanded (95% confidence limits) the range of variability increases. Because deer density estimates are relatively crude a rather low level of precision can be accepted and for this reason 80% confidence limits can be used. If the range is unacceptably large, further stratification of the habitat type should reduce the variation and improve the estimate.

Several dung counts for each habitat type should be obtained and care must be taken that values derived in one area are not inappropriately applied in another.

Vantage point counts

In hilly terrain, sample counts of deer over 2–3 h periods from a vantage point can provide useful support to dung counting. Vantage point counts are not so useful as dung counts because the count only relates to the number of deer using the area at the time of the count.

Counts in all structural types present should be conducted and count areas should be representative of the surrounding habitat. Counting deer emerging on to an area with young trees from surrounding thicket or pre-felling age crops is not recommended as it is impossible to relate the deer seen to a finite area. April and May are most suitable and periods from 06.00–10.00 and 16.00–21.00 span the best times of day for observing, especially when human disturbance is a problem. During each count, areas of 40–100 ha are methodically scanned in order to locate deer. Binoculars (7–10 × 50 mm) and a telescope (15–60 × 60 mm) are used for location and classification into sex and age classes, respectively, and are satisfactory at ranges up to about 1 km. Deer occupying even the densest habitats can be seen, though often only for short periods of time as they cross areas of broken canopy. Watches of more than 2 h duration are needed to allow sufficient time for the completion of rumination cycles and the start of a new feeding bout, which is when animals are most likely to be seen. Thus, by the time the count is completed, all deer present at the beginning of the observation period will have a high probability of being counted.

All sitings of deer are recorded, at the time, on a large scale map, with the time and direction of movement and also on a form (Figure 5). In the event of groups disappearing into thick cover, new groups are only added to the count if the time of emergence and the group composition suggests beyond all reasonable doubt that none of the animals has been previously counted. Deer moving into the area during the count period should not be counted.

At the end of the count period the minimum number of deer is related to the area under observation and expressed as a density per square kilometre (n km^{-2}). Three or four repeat counts are made over the following mornings and evenings and fresh counts in adjacent areas of the same, and other, structural types are made. Repeat counts are necessary to reduce the chance of counts reflecting inactivity due to weather conditions and the maximum number recorded in any single count period should be accepted as this is likely to reflect actual numbers in a sedentary species such as roe deer. Mean densities for each structural type can then be applied to the area as a whole in relation to the component structural types present.

Predicting changes in forest structure

Having estimated deer densities related to available habitats it is now possible to predict changes in these habitats and consequently their changing capacity to support particular deer densities.

Growth rates and volume increments for commercial tree species can be predicted from age and top height using the yield class system. Yield classes can be expressed by a series of growth curves which predict the volume production for a particular species, given information on age and top height. Graphs can thus be constructed relating age to sequential growth stages, which relate to successional changes

DENSITY ESTIMATION FOR WOODLAND ROE DEER OCCUPYING HILLY TERRAIN

Forest: Carrick Conservancy: South Scotland Observer(s): J. McLean

Location: Sron cruicte Area: 110 ha Structural Type: Thicket P Yr: 73

Date: 16·4·91 Time on: 06:30 Time Off: 10:00

Weather Conditions and Visibility: Fine and clear following a frost.

DEER SEEN

Time	Location	Spp	Adult Male	Adult Female	Yearlings Male	Yearlings Female	Calf/Kid	Unclassified	Remarks
0710	Ride/river	Roe		1		1	2		cumulative Total 4 Roe
0722	Check area in Compt 86	Red		1			1		Cum Total 4 Roe 2 Red
0730	Riverside	Roe		1		1	2		same as 07·10
0733	"	"	1						cum Total 5 Roe 2 Red
0752	Ride 86/87	Red		1		1	1		" " 5 Roe 5 Red
0804	Check on edge of 87 (N)	Roe		1			1		possibly part of 07·10 group cum Total 5 Roe 5 Red
0815	Check 87	Roe	1		1				cum Total 7 Roe
0820	Riverside	Roe	1						same as 0733
"	Check 87			1			1		" " 0804
0840	Ride 87/88	Roe		1			2		cum Total 10 Roe 5 Red
0845									
1000	– Continued sightings of 0710 group plus intermittent views of others								

No new sightings confirmed.

∴ 10 Roe deer

≏ 9·1/100 ha

Figure 5. Form for recording vantage point counts.

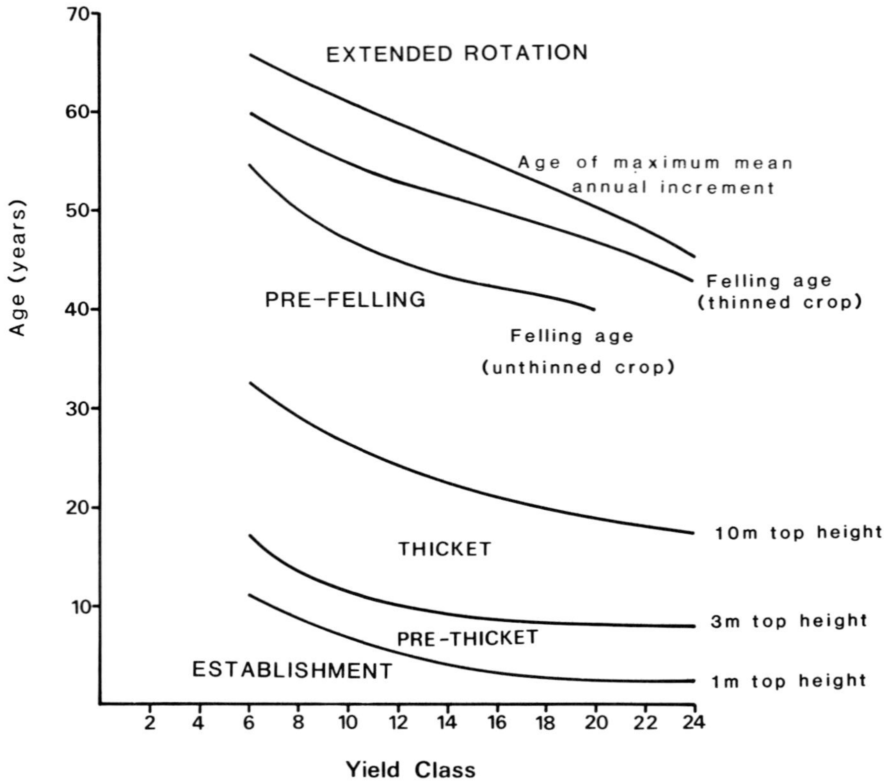

Figure 6. Model of structural change through a forest rotation.

affecting the deer (Figure 6). The areas occupied by each stage at any particular time can be computed, based upon a knowledge of the yield class.

The predicted changes in habitat availability at Alice Holt Forest are shown in Figure 7. More value would be gained from plotting curves for all structural types used for dung counts. Roe deer reach high densities in the early growth stages, and therefore the forest will be able to support considerably more deer between about 1995 and 2000, when there is an abundance of establishment and pre-thicket. The rapid onset of canopy closure as trees enter the thicket stage after 1998, will be accompanied by a reduction in pre-thicket. This suggests a marked decline in the capacity of the habitat to support deer, and an increasing impact on the habitat and the tree crop can be expected as high numbers of deer compete for diminishing resources. Culling can be planned in anticipation of these changes. The predictions assume the felling of crops at the age providing maximum net discounted revenue at a 5% discount rate, and replanting in the following year.

Population dynamics

Changes in the deer population depend upon natality, mortality, immigration and emigration. In most roe populations the number of yearlings attempting to establish themselves in the population is too great for the resources and space available. Consequently many yearlings, particularly males, emigrate and/or die. Their fate will depend upon whether vacant niches are available. Movements of deer resulting in a net loss to the population being managed can be considered as mortality.

Highest mortality occurs in the first year of life and this can be assessed indirectly by comparing estimates of pregnancy made during the female shooting season with kid:doe ratios determined from the observation of groups of deer

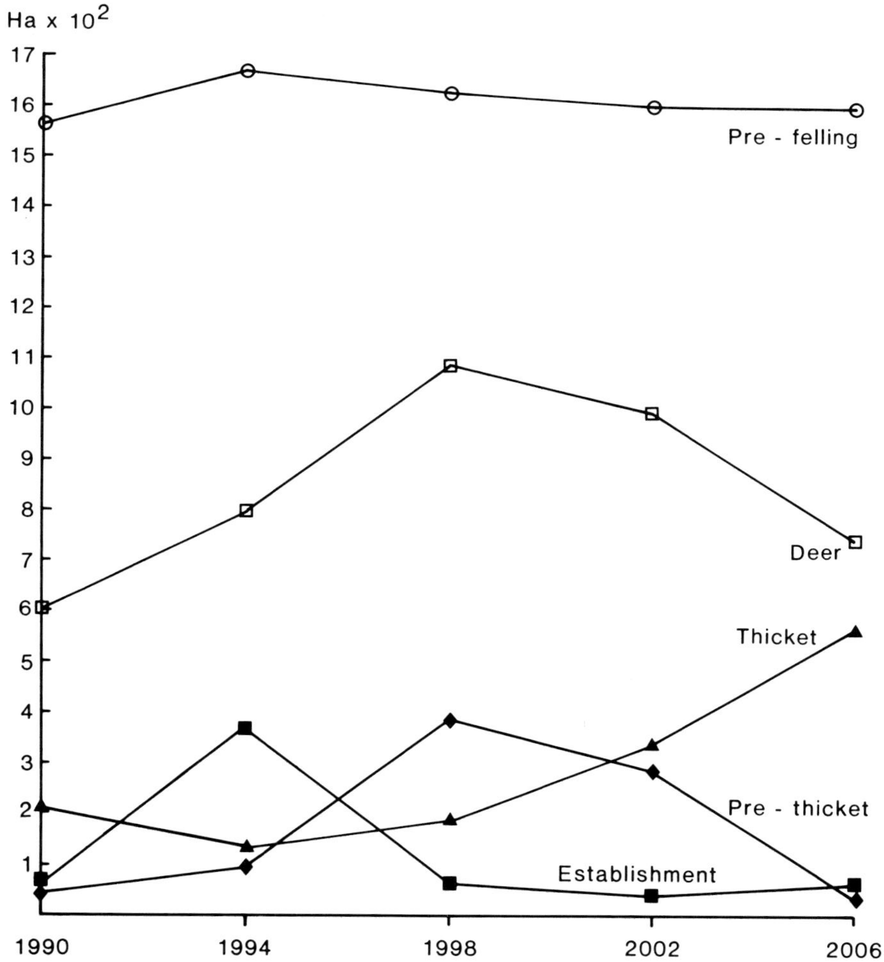

Figure 7. Predicted changes in forest structure/habitat availability for roe deer at Alice Holt Forest 1990-2006.

seen during autumn and early winter. This linking of reproduction with mortality provides a measure of kid survival and an estimate of the number of animals being recruited into the population.

The age of deer cannot be reliably determined from body weight, appearance, behaviour or antler size and shape, and in assessing population dynamics the age of all deer should be assessed post-mortem, including those from culling, accidents and natural mortality. This can only be determined from an examination of the lower jaw. The age of female deer must be related to their reproductive status and these data recorded. Ages of both sexes should be recorded for subsequent use in cohort analyses as described later. All lower jaws should be indelibly labelled and retained for age-determination at the end of the shooting season (see Appendix).

Reproductive performance does not usually vary greatly in adult deer, but the yearling and immature age classes are sensitive to changes in the relationship between density and habitat quality that influence reproductive potential. It is valuable, therefore, to separate kids from older deer by the presence or absence of milk teeth (see Appendix).

Assessing fertility

After shooting, bleeding and eviscerating, the

lower jaw should be removed. The reproductive system from all females (uterus, oviducts and ovaries) should be removed and the uterus cut open and examined (Plate 1). In the early stages of pregnancy up to the time of implantation in January, before any embryo is visible, it is easier to detect pregnancy by examining the ovary for the presence of corpora lutea. The corpus luteum is a yellowish gland which develops inside the ovary and only persists following fertilisation. A corpus luteum results from the release of each fertilised egg from the ovary. The number of corpora lutea therefore reflect the potential number of offspring per doe. The corpora lutea are easy to see and count if the ovary is sliced two or three times longitudinally (Plate 2).

The deer should be given a unique reference number which should be recorded in the ranger's or stalker's notebook and also on a label (indelible) which is fixed with wire to the lower jaw. The presence and number of corpora lutea and embryos should be recorded in the notebook alongside the reference number, date, location and weight (Figure 8). Labelled jaw bones should be retained until ages have been determined for cohort analyses.

An example of fertility measured in seven populations is given in Figure 9. This shows a mean number of corpora lutea found in kids, yearlings and adults. At Pickering (North Yorkshire), Alice Holt (Hampshire) and Thetford (Suffolk/Norfolk) the mean number of corpora lutea for adult does is about 2.0, while at Queens (Highland Region) the mean is only just above 1. This suggests that most adult does at Pickering, Alice Holt and Thetford have the potential to produce twin kids (high performance), while those in Queens Forest will only have single births (low performance).

Kids and yearlings are more sensitive than adults to food shortage, lack of shelter and inclement weather, which may result in poor

DEER CULL DATA SHEET

FOREST / DISTRICT West Downs, Alice Holt

Deer No.	Date shot	Location comp	Sex	Jaw collected	Embryos: Present	Embryos: No.	Embryos: Sex	Ovaries: check	Ovaries: Corpus luteum No.	Mam gland milk present	Carcass weight	Comments
8	17·12·86	comp 7	F	✓	–	–	–	✓	1+1=2	✓	16·0 kg	dog Tapeworm Cyst
9	17·12·86	7	F	✓	–	–	–	✓	1+1=2	✓	17·75	
10	17·12·86	7 Straits	F	✓	–	–	–	✓	0+0=0	–	10·0	
11	9·1·87	12	F	✓	–	–	–	✓	2+0=2	–	15·0	
12	9·1·87	71	F	✓	✓	2	? ?	✓	2+0=2	–	15·0	
13	9·1·87	71	F	✓	–	–	–	✓	1+0=1	–	15·0	
14	26·1·87	12	F	✓	✓	2	♂+♂	✓	2+0=2	–	13·0	
15	26·1·87	12	F	✓	✓	2	?+?	✓	1+0=1	–	15·0	
16	26·1·87	12	F	✓	–	–	–	✓	0+0=0	–	9·0	
17	26·1·87	13	F	✓	–	–	–	✓	0+0=0	–	9·0	
18	27·1·87	68	F	✓	–	–	–	✓	0+0=0	–	10·0	

Figure 8. Specimen deer cull entry in a ranger's notebook.

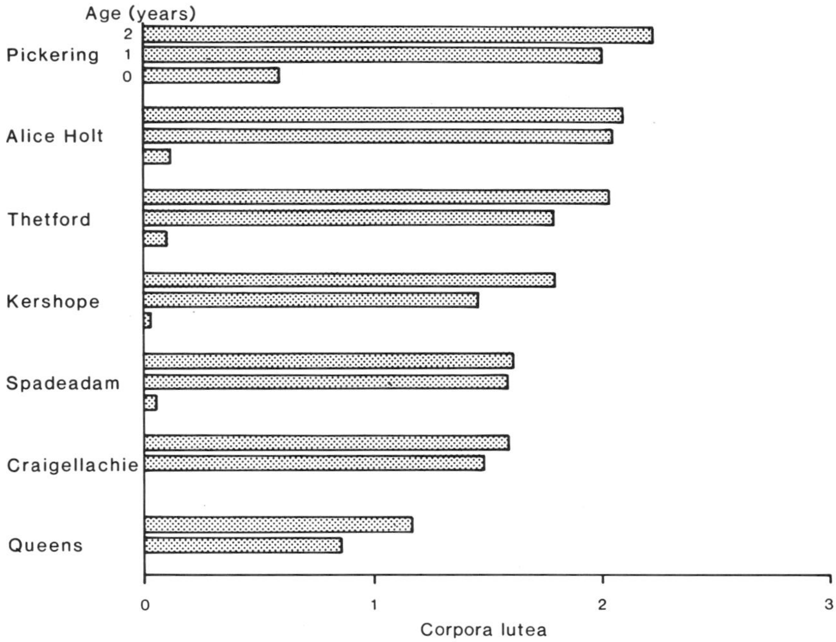

Figure 9. Mean number of corpora lutea for three age classes of roe deer.

Table 2. Percentage of yearling roe does carrying 0, 1, 2, 3, 4, corpora lutea.

Forest	*Corpora lutea* 0 % (*n*)	1 % (*n*)	2 % (*n*)	3 % (*n*)	4 % (*n*)	Total *n*
Pickering		4.0(1)	92.0(23)	4.0(1)		25
Alice Holt		4.8(2)	85.7(36)	9.5(4)		42
Thetford	1.9(3)	21.8(34)	72.4(113)	3.2(5)	0.6(1)	156
Kershope	4.5(1)	45.5(10)	50.0(11)			22
Spadeadam		41.7(5)	58.3(7)			12
Craigellachie	3.7(1)	44.4(12)	51.9(14			27
Queens	14.3(1)	85.7(6)				7

Table 3. Percentage of female roe kids carrying 0, 1, 2, 3 corpora lutea.

Forest	*Corpora lutea* 0 % (*n*)	1 % (*n*)	2 % (*n*)	3 % (*n*)	Fertile %	(*n*)	Total *n*
Pickering	52.0(39)	38.6(29)	8.0(6)	1.4(1)	48.0	(36)	75
Alice Holt	90.6(48)	7.5(4)	1.9(1)		9.4	(5)	53
Thetford	95.2(80)		4.8(4)		4.8	(4)	84
Kershope	97.7(42)	2.3(1)			2.3	(1)	43
Spadeadam	95.0(19)	5.0(1)			5.0	(1)	20
Craigellachie	100(32)				–	–	32
Queens	100(32)				–	–	32
					0.5	42	

growth rates, lower body weights and low rates of ovulation and conception. The proportion of yearling does with two or more corpora lutea is also greater in high performance populations (Table 2). Precocious reproduction (kid fertility) occurred mostly in the high performance populations, with 48% of kids from Pickering being fertile (Table 3). Does with the highest number of corpora lutea generally have higher body weights.

After mid January, embryos should be clearly visible on opening the uterus. Not all fertilised eggs will implant, and result in a viable embryo, so the potential reproductive capacity, assessed by the presence of corpora lutea may not be fully realised. The differences in fertility between populations, judged by a comparison of counts of corpora lutea (Figure 9), become reduced when judged by counts of embryos (Figure 10). Only does shot after 20th January are included. Comparison of Figures 9 and 10 shows that loss of potential embryos between fertilisation and implantation was consistently above 20% and reached 68% for adult does at Queens Forest. This is considerably greater than the 10–12% cited by Loudon (1980, unpublished) but is similar to those calculated by Chapman and Chapman (1971) and Chaplin (1977).

Only at Pickering were kids carrying embryos found, although kids with corpora lutea were found at Pickering, Alice Holt, Thetford, Kershope and Spadeadam, Northumberland. Sexually mature kids, and indeed some yearlings, may only achieve body weight consistent with ovulation later in the year than the usual rutting period. This may prolong rutting into the autumn, thus explaining the phenomena of the 'false rut'.

All pregnancy data including those based on counts of corpora lutea and embryos should be recorded. However, it is clear that counts based on corpora lutea are less reliable and will require some adjustment (see above and Figures 9 and 10).

Assessing mortality

Direct searches for deer carcasses are inefficient in dense forests and some indirect method for the estimation of mortality is required. Most adult deer that are shot during routine culling operations are in good condition with fat deposits around the kidneys, in the abdominal cavity and under the skin. In many forest populations

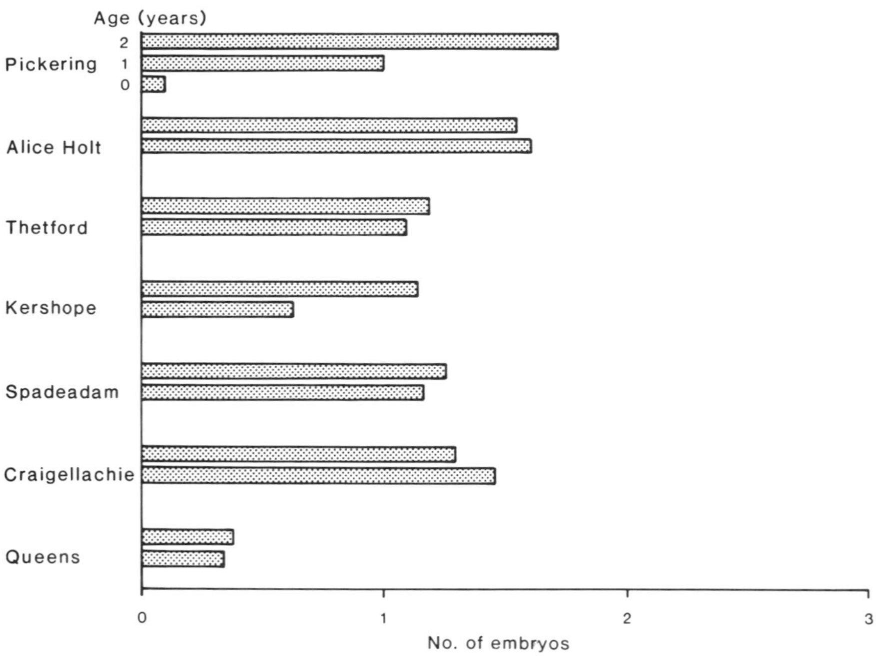

Figure 10. Mean number of embryos for three age classes of roe deer.

natural mortality is low in the middle age range but higher in juveniles and older animals. An examination of humerus or femur bone-marrow in adult animals found dead will indicate whether malnutrition was the cause of death (Ratcliffe, 1980). A red viscous marrow indicates almost total loss of fat due to malnutrition.

Late autumn and winter counts of groups of deer will indicate kid:doe ratios (Table 4), and this can be compared with pregnancy data to provide a measure of kid mortality (Table 5). However, it can be very difficult to distinguish juvenile roe from adults in the spring following their birth (age 8–10 months), and an assessment of winter and spring mortality by this means is unreliable. Detailed data on the survival of different age classes in an unshot population at Chedington Woods, Dorset are available, suggesting a mortality of about 50% of kids born within their first year of life (Gill, 1988, unpublished). Most of this mortality probably occurs during the first few weeks of life, and inclement weather, especially prolonged heavy rain, during the first few days of life is likely to increase kid mortality further (Plate 4). It seems likely that a high rate of density-dependent mortality in kids is common from year to year (*c.* 30%), but that density-independent mortality, caused by local weather conditions at the time of kidding, can significantly modify this. For example, there are very large differences in kid:doe ratios between years at Alice Holt, Kershope and Spadeadam forests, and there is no consistency within years (Table 4). In 1986 a very low kid:doe ratio was apparent at Alice Holt, while at Kershope and Spadeadam it was high. Kid survivorship, calculated by comparing potential birth rates (Figure 10) and kid:doe ratios (Table 4) ranged from 0.25 to 0.75 in the data available here (Table 5). The wide variability between years makes it impossible to predict this from year to year, and kid:doe ratios in autumn and winter should be used annually as the basis for estimating recruitment to the population, supported by data on fertility and pregnancy.

The considerable effect of inclement weather following kidding is very clearly demonstrated by relating the very wet periods in late May to early June of 1971 and 1981 at Alice Holt Forest (local meteorological data collected by the late Judith J. Rowe), to the recovery of animals in

Table 4. Roe deer kid:doe ratios calculated for deer observed in autumn-winter.

Forest	*Years*	*Does*	*Kids*	*Kids:doe*
Pickering	1985	70	39	0.56
Alice Holt	1985	16	10	0.625
	1986	16	6	0.375
	1987	10	7	0.7
Kershope	1985	56	18	0.32
	1986	49	33	0.75
Spadeadam	1986	56	47	0.84
	1988	35	20	0.57
Craigellachie	1987	64	28	0.44
	1986	58	21	0.36
Queens	1985	14	2	0.14
Affric	1985	17	4	0.23

Table 5. Roe deer kid survivorship.

Forest	*1* *Potential birth rate** *(kids:doe)* *(Figure 10)*	*2* *Autumn–Winter* *(kids:doe)* *(from Table 4)*	*3* *Kid survivorship* *to age 1* *Col. 2 ÷ Col.1*
Pickering	1.4	0.56	0.4
Alice Holt	1.5	0.38–0.7	0.25–0.47
Kershope	1.0	0.32–0.75	0.32–0.75
Spadeadam	1.2	0.57–0.84	0.48–0.7
Craigellachie	1.35	0.36–0.44	0.27–0.33
Queens	0.35	0.14	0.4

* Potential birth rates are derived, in this example, from an approximate mean of the number of embryos per 1 and 2 year old doe from Figure 10.

cohort analyses (Figure 13). It is clear that very few animals survived from these particular cohorts, only seven and five animals, respectively, being recovered.

Modelling population changes

Having estimated deer numbers, recruitment and mortality, this information can be used to model the predicted changes in the population and to set the cull relative to the management objectives. Population modelling can be done more quickly and probably more accurately with a computer, but can also be done by hand.

Models have been constructed for a range of populations from which reliable data are available, and for which additional data can be realistically estimated (Tables 6a–d). The proportion of deer in each age class has been derived from those determined from a detailed study at Chedington Woods, Dorset, and modified by local data on culls, fertility and survivorship.

The most detailed data for any roe deer population are available for Chedington Woods, Dorset. This population occupies a very rich lowland woodland and has not been shot for about 25 years. A simulation of this population suggests that in spite of no shooting and high natural mortality it will increase by 180% over 10 years (Table 6a). This population has probably reached its ecological carrying capacity and in reality is now regulating its own size by emigration as well as through reduced fecundity and high mortality.

When populations are culled natural mortality will usually decline; in other words, at least some of the animals that are shot will be those which would have died naturally. Therefore, when we attempt to simulate changes in shot populations we must increase the survivorship values based on unshot populations. At Chedington, this resulted in a 0.21 (21%) cull being necessary to achieve zero population growth (Table 6a).

Most other populations studied are culled regularly and realistic, but assumed, survivorships have been applied. At Alice Holt, Hampshire, a 0.25 (25%) cull achieved zero population growth (Table 6b), while at Kershope only a 0.17 (17%) cull was required (Table 6c). At Pickering a 0.26 (26%) cull was required (Table 6d). These cull levels are surprisingly low compared with recent suggestions that culls in the order of 30–40% were necessary to achieve zero population growth in roe deer (Loudon, 1980 (unpublished); 1982). Furthermore, if these models are modified by using the lower values of kid survivorship calculated from doe:kid ratios (Table 5), then the population at Kershope remains stable in the absence of culling. The reason for this appears to be that in reality very low survivorship occurs only infrequently, when very wet and cold conditions prevail immediately following birth (i.e. 1971 and 1981 at Alice Holt; see 'Assessing mortality' pp. 16-17). It is also clear that small changes in kid and year-

Table 6a. Simulations of roe deer population changes. Input data for Chedington Woods, Dorset.

		Kid	*Yearling*	*2–6 yr*	*7–9 yr*	*>10 yr*
Proportion of population	M	0.172	0.094	0.201	0.019	0.001
in each age class	F	0.172	0.102	0.217	0.021	0.001
Fecundity (embryos/F)		0	0.88	1.5	1.0	1.0
Survivorship (unshot)	M	0.52	0.86	0.8	0.35	0
	F	0.57	0.78	0.8	0.7	0
The population increases by 180% over 10 years						
Survivorship (assumed	M	0.7	0.9	0.95	0.30	0
in presence of shooting)	F	0.75	0.9	0.95	0.30	0

If a cull is imposed on the population, survivorship increases and a cull of 0.21 is required to achieve zero population growth.

Table 6b. Simulations of roe deer population changes.
Input data for Alice Holt Forest, Hampshire.

		Kid	*Yearling*	*2–6 yr*	*7–9 yr*	*>10 yr*
Proportion of population in each age class	M	0.188	0.098	0.185	0.014	0.001
	F	0.188	0.106	0.204	0.015	0.001
Proportion in each age class (reflecting a 2:1 sex ratio)	M	0.14	0.068	0.107	0.003	0.001
	F	0.283	0.146	0.232	0.014	0.001
Fecundity (embryos/F)		0	1.6	1.54	1.54	1.54
Survivorship (in presence of shooting)	M	0.7	0.9	0.95	0.30	0
	F	0.75	0.9	0.95	0.30	0

Cull to achieve zero population growth (1:1 sex ratio) = 0.25
Cull to achieve zero population growth (2:1 sex ratio) = 0.31

Table 6c. Simulations of roe deer population changes.
Input data for Kershope, Cumbria.

		Kid	*Yearling*	*2–6 yr*	*7–9 yr*	*>10 yr*
Proportion of population in each age class	M	0.152	0.089	0.218	0.027	0.001
	F	0.152	0.096	0.235	0.029	0.001
Fecundity (embryos/F)		0	0.63	1.14	1.14	1.14
Survivorship (in presence of shooting)	M	0.7	0.9	0.95	0.30	0
	F	0.75	0.9	0.95	0.30	0

Cull to achieve zero population growth = 0.17

Table 6d. Simulations of roe deer population changes.
Input data for Pickering, N. Yorkshire.

		Kid	*Yearling*	*2-6 yr*	*7-9 yr*	*>10 yr*
Proportion of population in each age class	M	0.181	0.101	0.194	0.015	0.001
	F	0.181	0.108	0.201	0.017	0.001
Fecundity (embryos/F)		0	1.0	1.71	1.71	1.71
Survivorship (in presence of shooting)	M	0.75	0.9	0.95	0.30	0
	F	0.8	0.9	0.95	0.30	0

Cull to achieve zero population growth = 0.26
Increase fecundity of yearlings to 1.6 = culls 0.28
Decrease survivorship of kids to 0.4 cull = 0.12

ling survivorship will have considerable effect on population size as populations are composed of large numbers of these age classes. Clearly it is important to collect precise data on survivorship of kids from kid:doe ratios, as this will strongly influence the next year's cull targets.

The adult sex ratio will influence the results from simulations and in all those cited so far a 1:1 sex ratio has been assumed. However, in three unshot roe deer populations: Kalo, Denmark (Strandgaard, 1972), Chedington, Dorset, and Glentress, Peebleshire (Loudon, 1978), sex ratios of 2 females: 1 male have been reported. It is not clear whether non-territorial males or indeed any non-established animals have been included in these ratios, and a recent analysis of the Chedington population data (Gill, 1988, unpublished) did not support a sex ratio significantly different from 1:1 (Table 6a). However, if the proportion in each age class is redistributed to represent a 2:1 sex ratio, further simulations show little increase in the cull required to achieve zero population growth. At Alice Holt (Table 6b) the cull requirement is raised from 25% to 31%.

It seems clear that British roe deer populations are seldom capable of sustaining the high levels of cull which have hitherto been assumed (30–40%), and that 15–25% seems more usual. In terms of the management of roe deer, this may be of considerable importance, as it now seems much more likely that roe deer populations can be regulated by shooting, although current culling levels may often be below even the 15–25% required to reduce population size.

Setting the cull

The number to cull each year should be based firmly on a knowledge of the population dynamics and estimated future population size. The target cull should be decided upon during the early summer following the assessment of spring densities and making use of autumn/winter kid:doe ratios for the estimation of recruitment supported by fertility and pregnancy data. In years of poor kid survival (e.g. 1971 and 1981 at Alice Holt), the cull targets can be relaxed or the advantages can be used to reduce population size, depending upon management objectives. The cull target should be sub-divided into numbers of males, females and kids based upon their estimated proportions in the population. It is important that sufficient effort is devoted to breeding-age females, as this group will most significantly influence future population trends. Clear objectives for the ranger or stalker must be stated. For example, if deer population control is considered to be the most important duty other less important duties should not be allowed to conflict with this.

In the initial period of data collection when insufficient information is available on which to base cull levels, it is possible to obtain some idea of the required cull from the relationships given in Figure 11, which provides guidance on prob-

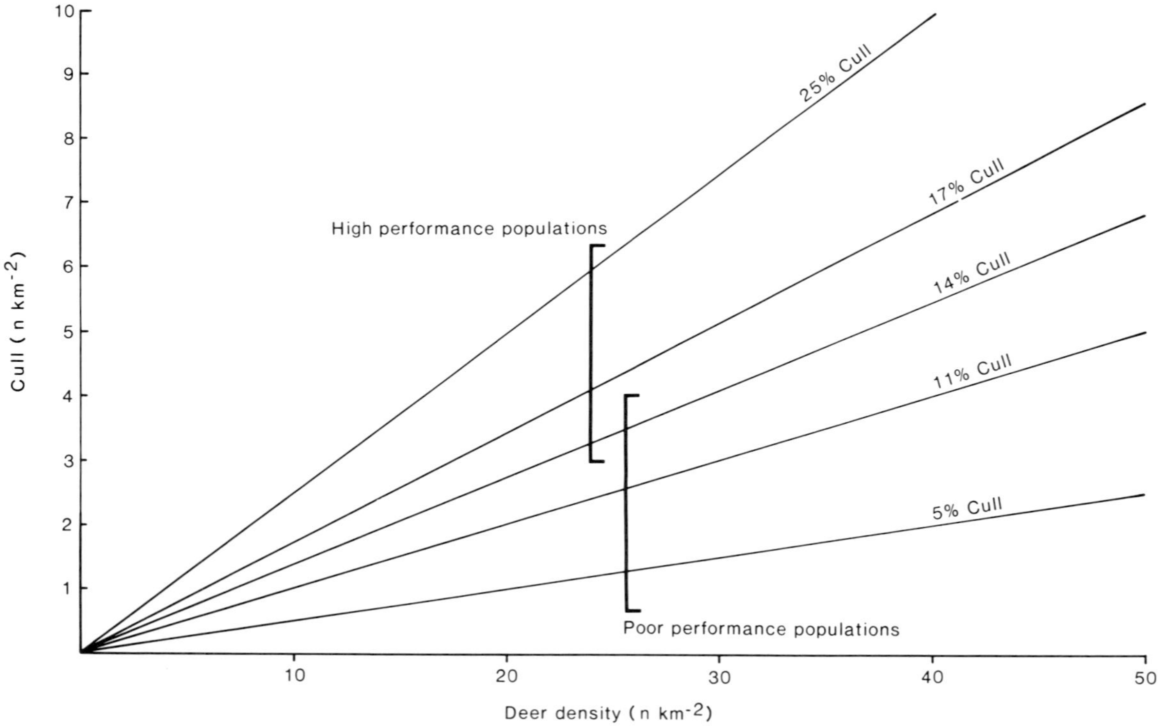

Figure 11. Cull requirements for roe deer populations.

Date	Nov. 5	6	7	8	10	11	12	13	16	17	18	19	20	23
Target		1		2		3		4		5		6		7
Kill							2 (2)			3 (5)				
Progress							–1			+				

Date	24	25	26	27	30	Dec. 1	2	3	4	7	8	9	10	11
Target		8		9		10		11		12		13		14
Kill									1 (6)	1 (7)		3 (10)		
Progress									–5	–5		–3		

Date	14	15	16	17	18	21	22	23	24	29	30	31	Jan. 4	5
Target		15		16		17		18		19		20		21
Kill	1 (11)	1 (12)		3 (15)									1 (16)	2 (18)
Progress	–3	–3		–1									–4	–3

Date	6	7	8	11	12	13	14	15	18	19	20	21	22	25
Target		22		23		24		25		26		27		28
Kill	3 (21)			3 (24)		1 (25)								
Progress	+			+1		+1								

Date	26	27	28	29	Feb. 1	2	3	4	5	8	9	10	11	12
Target		29		30		31		32		33		34		35
Kill					1 (26)									
Progress					–4									

Date	15	16	17	18	19	22	23	24	25	26	29
Target		36		37		38		39		40	
Kill	1 (27)		2 (29)	1 (30)						1 (31)	1 (32)
Progress	–8		–7	–7						–9	–8

Notes:
80 working days available.
Cull target = 40 breeding-age does.
Kids not included in record of progress.

Figure 12. Progress of roe doe cull, 1987/88.

able cull levels to achieve zero population growth for a range of densities. For example, a density of 20–30 deer km^{-2} in a high performance population will require a total annual cull of 3–7 deer km^{-2} to prevent an increase.

Monitoring the cull

Progress of culling should be carefully monitored preferably on a daily or at least on a weekly basis. A useful form for this purpose is shown in Figure 12. After deducting weekends and holidays, the number of days available for deer control can be shown against the planned cumulative cull for any point in time. In this way, problems (bad weather and illness preventing targets being achieved) are readily identified at an early stage and tactics can be modified accordingly. This form should also be used to analyse difficulties after completion of the cull in order to foresee and forestall similar problems in subsequent years.

Cohort analysis

A cohort is a collective name for all animals born in a single year. Cohort analysis is simply a means of recording and accumulating the number of deer born in particular years from a knowledge of their age at death. Clearly, it will be more accurate if non-cull deaths are recovered such as road casualties and natural mortality. It provides a retrospective check on minimum population size. This can be illustrated by maintaining a long-term record of the development of each cohort by annually adding cull data (Figure 13). By entering the number of animals belonging to each age class from the annual cull, in each vertical column, these can be added horizontally to give the cumulative

CULL YEAR

Cohort	70/71	71/72	72/73	73/74	74/75	75/76	76/77	77/78	78/79	79/80	80/81	81/82	82/83	83/84	84/85	85/86	86/87	87/88	Age	No. shot from cohort
70	6	3	5	2	0	0	3	0	1	0	0	0	0	0	0	0	0	0		20
71		0	1	3	2	1	0	0	0	0	0	0	0	0	0	0	0	0		7
72			4	6	2	2	2	0	1	0	1	0	0	0	0	0	0	0		18
73				2	2	1	4	5	1	1	0	0	1	0	0	0	0	0	15	17
74					7	4	2	2	2	1	0	0	0	0	0	1	0	0	14	19
75						4	5	4	6	5	1	0	0	1	0	0	0	0	13	26
76							5	3	10	3	2	0	0	0	0	0	0	0	12	23
77								4	4	7	3	0	0	0	0	0	0	0	11	18
78									13	2	7	3	0	0	0	0	1	0	10	26
79										9	3	3	4	2	1	0	1	0	9	23
80											8	8	2	1	4	1	0	0	8	22
81												2	1	1	0	1	0	0	7	5
82													3	3	6	6	2	1	6	21
83														1	0	0	1	1	5	3
84															5	3	4	2	4	14
85																5	4	1	3	10
86																	5	2	2	7
1987																		6	1	6
Total cull						12	21	18	38	28	25	16	11	9	16	17	18	13	0	
						+1		+2	+1	+2										

Figure 13. Roe deer cohort analysis, Alice Holt Forest.

number of deer shot for any particular cohort.

It is possible to estimate the minimum total population size in previous years by calculating the number of adults necessary to produce the number of kids born in a particular year. Cohort analysis will not start to yield useful data for 3–4 years, but it is a valuable deer management tool and it is strongly recommended that all the relevant data be collected from culls.

From the analysis at Alice Holt Forest (Figure 13) it can be seen, for example, that 26 deer have been recovered from the 1978 cohort. We know that about 0.7 kids/doe are recruited (survivorship in Table 6b), so approximately 37 mature does would have been present. Add to this, 37 males, about 20 yearlings, and the 26 kids which survived, and this brings the population to 120 deer in 1978, without including any natural deaths in the older age classes. Natural mortality in shot populations appears to be about 5% each year (Tables 6a–e), which is approximately equivalent to a further 40 deer over the 10-year life span of the cohort. So we can estimate that the total population in the winter of 1978 was about 160 deer (equivalent to 18 deer km^{-2}).

Of course cohort analyses assume that there have not been large numbers of deer losses or gains in the numbers of deer caused by immigration to, or emigration from, the area during the period of cohort reconstruction. This is unlikely to be a problem as gains will more or less equal losses unless the shooting strategies of neighbours are very different from your own.

There is a practical difficulty in applying cohort analysis to roe deer populations. The retrospective reconstruction of cohorts depends upon accurate age determination of all deer found dead or shot. Once roe deer have replaced the milk teeth at 10–15 months, it is only possible to determine their age by cutting the tooth in half and counting annual growth lines in the dental cementum microscopically (see Plate 3). However, although the method is tedious it is not difficult to apply (see Appendix).

Selective control

No mention has so far been made of selective control. Clearly, if sick or poor quality deer are present in a population then they should be culled. However, in many populations these are very few and the population must be controlled by shooting healthy, good quality animals, especially females. However, managers may wish to keep some good quality males. Because vacated territories are rapidly filled by non-territorial bucks and because territorial bucks can be difficult to identify there is no value in attempting to protect territorial or 'master' bucks, to reduce damage. The main criterion must always be the humane control of deer. Most care is necessary when out of season shooting before the female open season and during the early part of the season, when dependent offspring may be orphaned. In such cases every effort should be made to shoot the kid first.

Achieving the cull

It may become apparent that deer are breeding so well, and densities are so high, that cull targets are difficult if not impossible to achieve. This may be because increasing population size was not obvious in the past due perhaps to an increased area of pre-thicket woodland. Poor access is often the cause of failure to achieve the cull. Whatever the cause, the means must be found of achieving the cull targets required to meet management objectives.

Forest design is a vital part of deer management and the provision of open areas for the purpose of improving the possibilities for killing deer is essential. This aspect is covered more fully in Forestry Commission Leaflet 86 *Glades for deer control in upland forests* (Ratcliffe, 1985), but essentially a system of glades connected by forest roads and stalking paths should be maintained. Where possible, glades should be areas where the deer themselves have shown a preference for herb-rich swards and frequently these may be the areas of better quality land, but in some instances reseeding may be necessary. Deer glades are almost always complementary to conservation, streamside management, and landscaping objectives.

Forest rangers carrying out deer management must be skilled and well-equipped.

Training must be provided to ensure that rangers are competent in the biological principles of deer management and that they can use firearms safely and kill deer humanely. They should be equipped with rifles, binoculars and other equipment of a high quality appropriate to full-time professional use.

Transport and slaughtering facilities are also of great importance. Well-designed and equipped deer larders reduce physical effort and reduce loss of income by assisting good preparation of the carcass and allowing for improved operator hygiene.

In hilly terrain, high seats are often considered unnecessary but in the lowlands and in rolling terrain, they can be useful in providing an effective vantage point and a safe shooting position. The use of high seats is covered in Forestry Commission Leaflet 74 *High seats for deer management* (Rowe, 1979).

The flow chart (Figure 2) provides a management approach to decision making. If particular difficulties are foreseen, consideration should be given to extending seasons and/or night shooting where permissable. So long as the treatment of deer conforms to the same high standards as in-season and daytime shooting, then these methods provide an excellent means to supplement the cull. Careful trials and experience of night shooting in south Scotland show that this method is at least as humane and effective as daytime shooting. Before undertaking night shooting, personnel should be familiar with the legal constraints, particularly in England and Wales and the Red Deer Commission's code of practice.

Summary of management duties

1. Divide the forest area into sub-areas of discrete range, i.e. those containing separate populations, irrespective of administrative boundaries. Consult neighbours if necessary in order to set up deer management groups.
2. Where timber production is an objective, carry out regular damage inspections of all woodland blocks, followed by damage assessments (Forestry Commission Leaflet 82) where and when necessary.
3. Predict changes in forest and woodland structure (based upon woodland stock maps, yield class data and aerial photographs) and the future impacts of these changes on deer populations.
4. Conduct dung-counts in early spring. Defecation rates of 20 groups per day can be used. Decay rates should be determined locally for representative habitats. Supplement with vantage point counts in suitable hilly areas.
5. Collect jaw bones from all deer shot, for age determination and future cohort analysis. Attach a label with reference number to the jaw bone and retain all jaws. Record animal reference number, date shot, location, sex and weight.
6. For all female deer, assess fertility by examining the reproductive tract:
   a. How many foetuses are present?
   b. If no foetuses, how many corpora lutea are present?
   Record all this information with the animal reference number (paragraph 5 above).
7. Monitor culling achievement.
8. Record sex and age class for all groups of deer seen during September–January to obtain kid:doe ratios.
9. End of season (April–June).
   a. Determine ages from all jaw bones (Appendix) and relate to animal reference numbers.
   b. Summarise pregnancy (%) data for kids, yearlings and adults (see paragraphs 5 and 6 above).
   c. Summarise densities in different areas.
   d. Estimate kid mortality/survival by comparing autumn–winter kid:doe ratios with fertility rates.
   e. Estimate annual increase or decrease (allowing for the current cull).
   f. Complete cohort analysis record for the year and recalculate minimum population size for earlier years (see paragraph 9a above).
10. From information gathered estimate annual population changes and set next year's culls.

11. Consult the management strategy flow chart (Figure 2) to assist in decision making.
12. Address problems of access and forest design. Select sites for glades (Forestry Commission Leaflet 86).

ACKNOWLEDGEMENTS

We would like to thank all of the Forestry Commission rangers and Conservancy staff too numerous to mention by name, who have supported this work over many years. We feel especially grateful for the contribution made by the late Judith J. Rowe who, as Former Head of the Forestry Commission's Wildlife Research Branch, initiated much of the research on which this publication is based. We also thank Mr Andrew J. Peace for help with statistical analysis and interpretation, and Dr Brian Staines for his long-standing interest and encouragement. We are also grateful to Andrew Chadwick, Brian Staines and Richard Toleman for their constructive and critical comments.

REFERENCES

CHAPLIN, R. E. (1977). *Deer*. Blandford Press, Poole, Dorset.

CHAPMAN, D. I. and CHAPMAN, N. G. (1971). Further observations on the incidence of twins in roe deer, *Capreolus capreolus*. *Journal of Zoology, London* **165**, 505–544.

LOUDON, A. S. I. (1978). The control of roe deer populations: a problem in forest management. *Forestry* **51**, 73–83.

LOUDON, A. S. I. (1982). Too many deer for the trees? *New Scientist* **93** (1297), 708–711.

MELVILLE, R. C., TEE, L. A. and RENNOLLS, K. (1983). *Assessment of wildlife damage in forests*. Forestry Commission Leaflet 82. HMSO, London.

MITCHELL, B., ROWE, J. J., RATCLIFFE, P. R. and HINGE, M. (1985). Defecation frequency in roe deer (*Capreolus capreolus*) in relation to the accumulation rates of faecal deposits. *Journal of Zoology, London (A)* **207**, 1–7.

PEPPER, H. W. (1992). *Forest fencing*. Forestry Commission Bulletin 102. HMSO, London.

PEPPER, H. W., ROWE, J. J. and TEE, L. A. (1985). *Individual tree protection*. Forestry Commission Arboricultural Leaflet 10. HMSO, London.

POTTER, M. J. (1991). *Treeshelters*. Forestry Commission Handbook 7. HMSO, London.

RATCLIFFE, P. R. (1980). Bone marrow fat as an indicator of condition in roe deer. *Acta Theriologica* **25, 26**, 333–340.

RATCLIFFE, P. R. (1985). *Glades for deer control in upland forests*. Forestry Commission Leaflet 86. HMSO, London.

ROWE, J. J. (1979). *High seats for deer management*. Forestry Commission Leaflet 74. HMSO, London.

STRANDGAARD, H. (1972). The roe deer (*Capreolus capreolus*) population at Kalo and the factors regulating its size. *Danish Review of Game Biology* **7** (1), 1–205.

Appendix

Age Determination of Roe Deer

Age determination of juveniles

In roe deer, eruption of all permanent teeth is complete by 10–15 months of age, and the presence of a complete adult dentition is a means of separating juveniles from adults. Up to about 3 months of age kids have no molars and only the incisors and three milk pre-molars are present on each side of the jaw. In the succeeding months, the molars progressively erupt as the jaw lengthens, and the milk pre-molars are shed and replaced by the permanent pre-molars. The third milk pre-molar functions like a molar in the young animal and is tricuspid, but is replaced by a bicuspid permanent tooth (Figure A1). This provides a useful means of identifying juveniles.

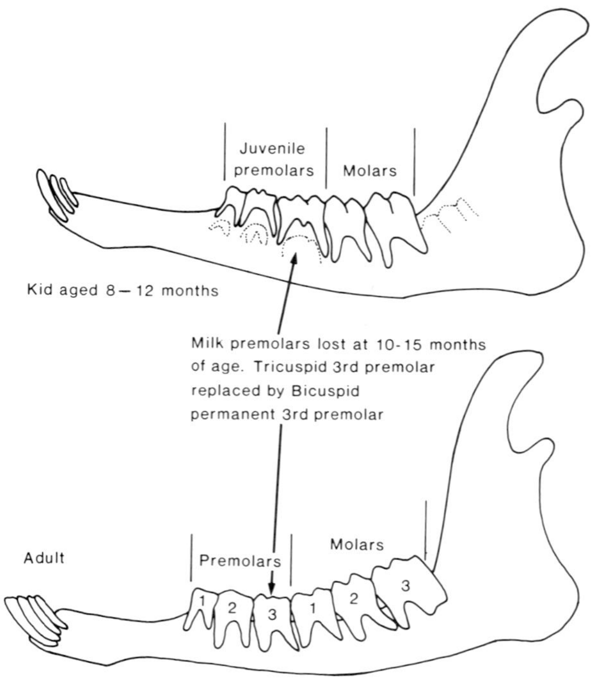

Figure A1. Juvenile and adult dentition.

Yearling bucks, shot around their first birthday in May–July, may have a complete adult dentition, but some of the milk pre-molars may be retained and 'sitting on top' of the permanent teeth beneath. The third molar may not be fully erupted (Figure A1). Yearling does shot during the winter will normally have complete adult dentition, with little wear evident on the third molar.

Precise assessment of the age of deer with a complete set of permanent teeth is only possible using the tooth sectioning technique (Aitken, 1975) described below.

Age determination of adults

Distinct layers of dental cement are laid down annually between the roots as the teeth grow. These annual layers can be counted with the aid of a microscope or powerful hand-lens to give the age of the animal. In roe deer, layers are clearest in the first and second molars. Broad white bands, composed of cellular cementum, represent the summer growth, while the narrow darker layers of acellular material represent the winter reduction in the rate of growth.

Method

1. Clean excess meat from jaw bone. Do not clean by boiling and do not use bleach, as this makes the cement layers brittle, making counting difficult or impossible.
2. Clamp the jaw bone in a vice. With a fine hacksaw cut the second molar into two pieces, X–X, cutting slightly behind the mid line, Y–Y (Figure A2), to at least 0.5 cm depth into the jaw.
3. Remove the front portion of the tooth. This may come out easily, if not, use a fretsaw to

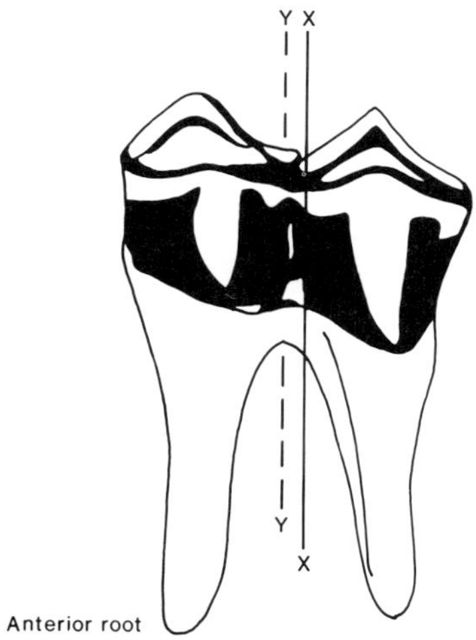

Figure A2. Cutting the first or second molar.
Line of cut X–X to the posterior side of the mid line (Y–Y).
Note: cutting should be done while the tooth remains within the jaw.

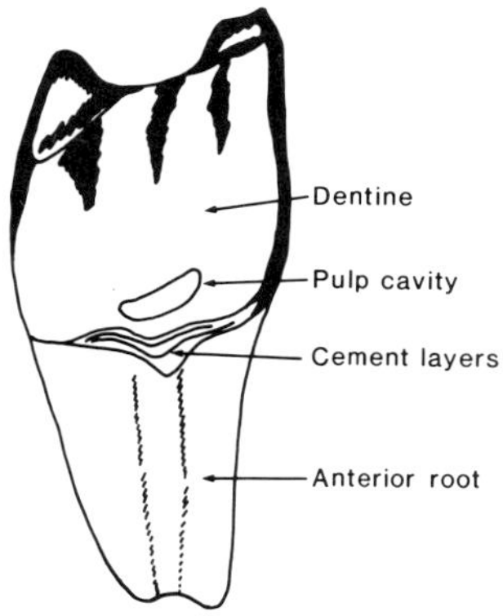

Figure A3. Transverse section through 1st or 2nd molar to show cementum pad location.

cut between the first and second molars and through the root of the tooth, below the level of the jaw bone, cutting in a line parallel with the line of the jaw, to ensure that the cement pad is not damaged.

4. Polish the cut face of the anterior portion of the tooth using fine emery paper (a fine carborundum stone may also be used for final polishing). Wipe the polished surface with a moist cloth or paper to remove any particles of grit that may be left.
5. Support the tooth, polished side uppermost, in Plasticine or Blu-tak on a small piece of plastic or wood that can be moved around on the microscope stage. Shine a light directly on the polished surface and view using a magnification of between ×10 and ×40.
6. Locate the cement pad (Figure A3, Plate 3) and count the wide white layers. Each white band represents one summer's growth, the first layer being laid down at around 1 year old.
7. In many cases the layers will be indistinct across some part of the cement pad. Repeat steps 4–6 to improve clarity so that the maximum number of layers can be counted. This number is equal to the animal's age in years.
8. If further polishing does not result in a clearer identification of the cementum layers the process should be repeated on the other half of the tooth or on the second molar from the other side of the jaw.
9. If the date of death is known, the age can be assessed to the nearest month, assuming 1st June as a common birth date; i.e. a doe shot on 1st December with four white layers visible will be 4 years 6 months old (Figure A4).
10. In adult bucks, fine discontinuous lines may be found within the white layers; these are apparently formed during the rutting period and do not represent annual increments to the cementum.
11. For animals shot during the summer, confusion may arise in assessing their age, as the current summer's layer of cellular cemen-

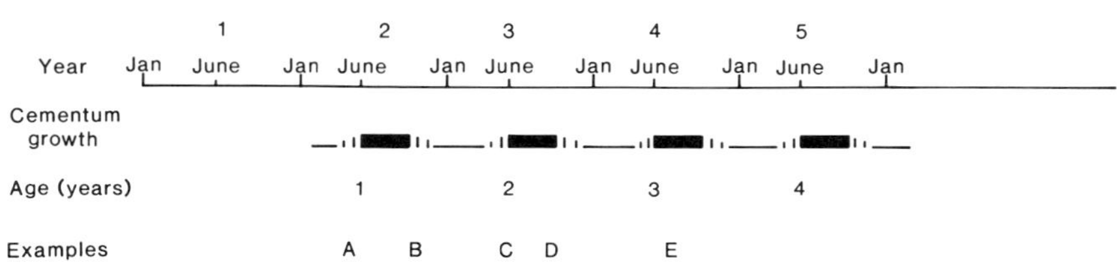

Figure A4. Deposition of annual cementum layers in relationship to cull seasons and age determination.

tum may or may not be visible. Before 1st June, the layer at the extreme lower edge of the cementum will normally be a narrow dark layer of acellular cementum as the current summer's white layer has not yet formed. The age of the animal is calculated by counting the number of white layers plus months from the previous 1st June. For example, an animal shot at the end of March showing three white layers will be 3 years and 10 months. After 1st June, if the layer at the extreme lower edge is a white layer of cellular cementum (the current summer's layer), the age is calculated by counting white layers and adding months from 1st June to date of death. If the layer at the extreme lower edge is a narrow, dark layer of acellular cementum in a deer dying after 1st June, then 1 year should be added to the calculated age as growth of the current summer's white layer has not occurred but the animal has obviously passed its current birthday.

The formation of distinct layers within the cementum is dependent upon there being a difference in the growth rates of animals between winter and summer. In some populations, in good quality habitats with less extreme weather conditions, there may be little difference between summer and winter growth rates, and so the layers in the cementum may be less distinct.

REFERENCE

AITKEN, R. J. (1975). Cementum layers and tooth wear as criteria for ageing roe deer (*Capreolus capreolus*). *Journal of Zoology, London* **175**, 15–28.

Printed in the United Kingdom for HMSO
Dd.293241 C30 6/92 (PP contract number 518) (CCN 51306)